Collins

KS2 Maths

W0013462

Maths

Age 10 – 11

Key Stage 2

SATs
Practice
Test Papers

2 × tests

SATs
Practice
Test
Papers

✓ School pack

Collins KS2 Maths

Prepare children for KS2 SATs with confidence

Revision Guides
Clear and accessible revision with five spaced practice opportunities for every topic.

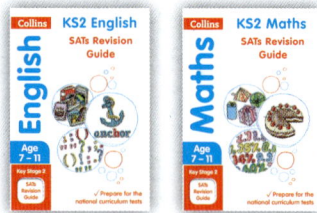

Practice Workbooks
Practice questions for every KS2 topic with short tests to assess progress.

Targeted Practice Workbooks
Focused year-by-year practice to target every topic.

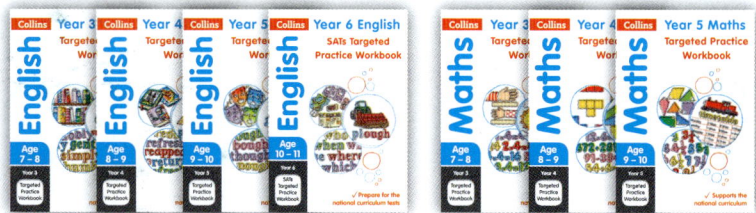

SATs Question Books
SATs-style questions for realistic test preparation.

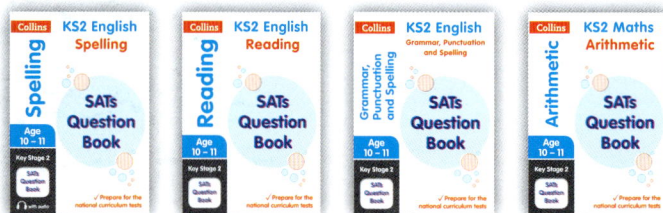

Practice Test Papers
Actual test-style papers for familiarisation.

For more information visit

collins.co.uk/collinsks2revision

Key Stage 2 Maths Practice Test Papers
Contents, Instructions and Answers

Author: Faisal Nasim

Contents

Set A

Mathematics Paper 1: arithmetic

Mathematics Paper 2: reasoning

Mathematics Paper 3: reasoning

Contents, Instructions and Answers

Set B

Mathematics Paper 1: arithmetic

Mathematics Paper 2: reasoning

Mathematics Paper 3: reasoning

Introduction

This practice resource consists of two complete sets of Key Stage 2 maths practice test papers. Each set contains similar test papers to those that pupils will take at the end of Year 6. They can be used any time throughout the year to provide practice for the Key Stage 2 tests.

The result of the papers will provide a good idea of pupils' strengths and weaknesses.

Administering the Tests

- Children should work in a quiet environment where they can complete each test undisturbed.
- Children should have a pen or pencil, ruler, eraser and protractor. A calculator is not allowed.
- Handwriting is not assessed but children should write their answers clearly.
- The amount of time allowed per test varies, so children should check the time given on each test paper.

Marking the Tests

Each set of maths practice papers is worth a total of 110 marks:
- Paper 1 Arithmetic is worth 40 marks
- Paper 2 Reasoning is worth 35 marks
- Paper 3 Reasoning is worth 35 marks

Use this answer booklet to mark the test papers and add up the total marks for Paper 1, Paper 2 and Paper 3. As a general guideline, if a child gets 58 or more marks across the three papers (i.e. 58 or more out of 110), they are reaching the expected standard. Keep in mind that the exact number of marks required to achieve the expected standard may vary year by year depending on the overall difficulty of the test.

Acknowledgements

The author and publisher are grateful to the copyright holders for permission to use quoted materials and images.

All illustrations © HarperCollins*Publishers*

Every effort has been made to trace copyright holders and obtain their permission for the use of copyright material. The author and publisher will gladly receive information enabling them to rectify any error or omission in subsequent editions. All facts are correct at time of going to press.

Published by Collins
An imprint of HarperCollins*Publishers*
1 London Bridge Street
London SE1 9GF

© HarperCollins*Publishers* Limited 2018
ISBN 9780008278168
First published 2018
10 9 8 7 6 5 4 3 2 1

All rights reserved. No part of this publication may be reproduced, stored in a retrieval system, or transmitted, in any form or by any means, electronic, mechanical, photocopying, recording or otherwise, without the prior permission of Collins.

British Library Cataloguing in Publication Data.
A CIP record of this book is available from the British Library.

Author: Faisal Nasim
Commissioning Editor: Michelle I'Anson
Editor and Project Manager: Katie Galloway
Cover and Inside Concept Design: Paul Oates
Text Design and Layout: Aptara® Inc
Production: Lyndsey Rogers
Printed in the UK by Martins The Printers

Answers

Content domain coverage for the questions in this paper are shown in the tables of answers below. Information about these codes can be found in the KS2 Maths test framework.

Set A Paper 1: arithmetic

Question (Content domain)		Requirement	Mark
1	(4N2b)	3,023	1
2	(4C2)	3,203	1
3	(4F4)	$1\frac{3}{8}$ OR $\frac{11}{8}$ Accept equivalent mixed numbers, fractions or exact decimal equivalent, e.g. 1.375 Do not accept rounded or truncated decimals.	1
4	(4C6b)	345	1
5	(3C1)	385	1
6	(5F8)	9.523	1
7	(4C2)	4,500	1
8	(3C7)	182	1
9	(4C6a)	7	1
10	(4C7)	705	1
11	(4C2)	3,166	1
12	(4F4)	$\frac{7}{20}$ Accept equivalent fraction, e.g. $\frac{28}{80}$ or **exact** decimal equivalent, e.g. 0.35	1
13	(3N2b)	2,289	1
14	(6C9)	78	1
15	(6F5a)	$\frac{1}{7}$ Accept equivalent fraction, e.g. $\frac{6}{42}$ or **exact** decimal equivalent, e.g. 0.14	1
16	(5C6a)	3,000	1
17	(5C7b)	59	1
18	(6F9a)	0.008	1
19	(5C6b)	3,465,000	1
20	(6C7b)	Award **TWO** marks for answer of 32 Working must be carried through to a final answer for **ONE** mark. Award **ONE** mark for a formal method of division with no more than **ONE** arithmetic error.	Up to 2
21	(4F8)	5.36	1
22	(6C7a)	Award **TWO** marks for answer of 193,500 Award **ONE** mark for a formal method of long multiplication with no more than **ONE** arithmetic error. Working must be carried through to a final answer for **ONE** mark. **Do not** award any marks if the error is in the place value, e.g. the omission of the zero when multiplying by tens.	Up to 2

Question (Content domain)		Requirement	Mark
23	(5F4)	$\frac{1}{10}$ Accept equivalent fractions or an **exact** decimal equivalent, e.g. 0.1	1
24	(6C7a)	Award **TWO** marks for answer of 32,674 Award **ONE** mark for a formal method of long multiplication with no more than **ONE** arithmetic error. Working must be carried through to a final answer for **ONE** mark. **Do not** award any marks if the error is in the place value, e.g. the omission of the zero when multiplying by tens.	Up to 2
25	(5F8)	22.161	1
26	(6F4)	$\frac{3}{5}$ Accept equivalent fraction, or **exact** decimal equivalent, e.g. 0.6	1
27	(6F5b)	$\frac{1}{8}$ Accept equivalent fraction, e.g. $\frac{3}{24}$ or **exact** decimal equivalent, e.g. 0.125	1
28	(6F5b)	$\frac{5}{12}$ Accept equivalent fractions or **exact** decimal equivalent, e.g. 0.416	1
29	(6R2)	182 Do not accept 182%	1
30	(6F4)	$4\frac{1}{10}$ OR $\frac{41}{10}$ Accept equivalent mixed numbers, fractions or **exact** decimal equivalent, e.g. 4.1 Do not accept rounded or truncated decimals. Do not accept $3\frac{11}{10}$	1
31	(6R2)	56 Do not accept 56%	1
32	(6F4)	$\frac{11}{20}$ Accept equivalent fractions or **exact** decimal equivalent, e.g. $\frac{22}{40}$ or 0.55	1
33	(6F9b)	180	1
34	(6R2)	240 Do not accept 240%	1
35	(5F5)	$94\frac{1}{2}$ Accept equivalent fractions or **exact** decimal equivalent, e.g. 94.5 or $\frac{189}{2}$	1
36	(6C7b)	Award **TWO** marks for answer of 39 Working must be carried through to a final answer for **ONE** mark. Award **ONE** mark for a formal method of division with no more than **ONE** arithmetic error.	Up to 2

Set A, Paper 2: reasoning

Question (Content domain)		Requirement	Mark	Additional guidance
1	*(4S2)*	190 10	1 1	
2	*(4C6a)*	Three boxes completed correctly as shown: | x	9	5
8	72	40
12	108	60		1	
3	*(5C6b)*	The correct number circled as shown: (4030)	1	Accept alternative unambiguous indications, e.g. number ticked.
4	*(4C4)* *(4S2)*	Award **TWO** marks for answer of 3,148 Award **ONE** mark for evidence of an appropriate method, e.g. • 2,693 + 1,012 = 3,705; 6,853 − 3,705	Up to 2	Answer need not be obtained for the award of **ONE** mark.
5	*(4M4c)*	Award **TWO** marks for boxes completed as shown: 6 years; 4 days; 11 weeks Award **ONE** mark for two boxes completed correctly.	Up to 2	
6	*(4N2b)*	Award **TWO** marks for boxes completed as shown: 	26	**1,026**
7,054	8,054
16,269	17,269	 Award **ONE** mark for two boxes completed correctly.	Up to 2	
7	*(5C4)*	Award **TWO** marks for answer of 2,884 Award **ONE** mark for evidence of an appropriate method, e.g. • 1,576 + 9,854 = 11,430; 11,430 − 8,546	Up to 2	Answer need not be obtained for the award of **ONE** mark.
8	*(4F2)*	(shapes ticked) ✓ ✓	1	Accept alternative unambiguous indications, e.g. shapes circled.
9	*(6A2)*	110 3	1 1	The answer is a time interval. Accept 1 hour 50 minutes.
10	*(5N4)*	Award **TWO** marks for boxes completed as shown: 76,620; 76,600; 77,000 Award **ONE** mark for two boxes completed correctly.	Up to 2	
11	*(6G2b)*	Award **TWO** marks for triangular prism and cube ticked. Award **ONE** mark for: • the cube and prism ticked and not more than one incorrect shape ticked OR • only one correct shape ticked and no incorrect shape ticked.	Up to 2	Accept alternative unambiguous indications, e.g. shapes circled.
12	*(5F8)*	0.428 0.72 2.134 2.5	1	
13	*(5M9a)*	Award **TWO** marks for answer of 91p Award **ONE** mark for evidence of an appropriate method, e.g. • 12 × 95p = £11.40; £11.40 − £10.49	Up to 2	Award **ONE** mark for answer of £91 OR £91p as evidence of an appropriate method. Answer need not be obtained for the award of **ONE** mark.
14	*(5N3b)*	2012	1	Do not accept answer in words.
15	*(6R4)*	20	1	Accept 20:15 OR 15:20
16	*(5G4b)*	(shape shown)	1	
17	*(6C8)* *(5M9a)*	Award **TWO** marks for answer of £1.73 Award **ONE** mark for evidence of an appropriate method, e.g. • 20 − 14.81 = 5.19; 5.19 ÷ 3	Up to 2	Award **ONE** mark for answer of £173 OR £173p as evidence of an appropriate method. Answer need not be obtained for the award of **ONE** mark.
18	*(6R2)*	15	1	

19	(6P3)	Quadrilateral completed as shown:	1	Accept very slight inaccuracies in drawing, within a 2 mm radius of the correct point.
20	(6F11)	An explanation showing that 0.4 is smaller than $\frac{4}{5}$, e.g. • $\frac{4}{5}$ is 0.8 > 0.4 • 0.4 is $\frac{4}{10} < \frac{8}{10}$ • 0.4 is 40% and $\frac{4}{5}$ is 80%. 40% is smaller than 80% • $\frac{4}{5} = 0.8$	1	Do not accept vague, incomplete or incorrect explanations, e.g. • Because $\frac{4}{10}$ is smaller than $\frac{4}{5}$ • Because $\frac{4}{10}$ comes first on a number line.

21	(5M9b)
	(6R3)
		Award **TWO** marks for answer of 15.5
		Award **ONE** mark for evidence of an appropriate method, e.g.
		• 775 ÷ 50
		OR
		• 100 km is 2 cm; 50 km is 1 cm
		  500 km is 10 cm; 25 km is 0.5 cm
		  10 cm + 2 cm + 2 cm + 1 cm + 0.5 cm
			Up to 2	Answer need not be obtained for the award of **ONE** mark.
				Do not accept incorrect proportions in any step without evidence of the calculation performed.

22	(6F4)	Award **TWO** marks for answer of $\frac{5}{8}$ Award **ONE** mark for evidence of an appropriate method, e.g. • $\frac{1}{4} + \frac{1}{8} =$ • $\frac{2}{8} + \frac{1}{8} = \frac{3}{8}$ • $1 - \frac{3}{8}$ • $\frac{1}{4} + \frac{1}{8} + \frac{1}{8} + \frac{1}{8}$	Up to 2	Accept equivalent fractions or exact decimal equivalent, e.g. 0.625 Answer need not be obtained for the award of **ONE** mark.
23	(6R3)	1:3	1	Accept other equivalent ratios, e.g. 3:9 Do not accept reversed ratios, e.g. 9:3

Set A, Paper 3: reasoning

Question (Content domain)		Requirement	Mark	Additional guidance
1	(5N2)	Award **ONE** mark for: E, B, C, A, D	1	Accept: £87,500 B £140,500 £147,250 £151,600
2	(3N2a)	698 771	1 1	

3	(3C2)	Award **TWO** marks for:
		  2 7 **4**
		+5 **5** 2
		  8 2 6
		Award **ONE** mark for two digits correct.
			Up to 2	

4	(5C5c)	Award **TWO** marks for all placed as shown: Award **ONE** mark for three numbers placed correctly.	Up to 2	Accept alternative unambiguous indications, e.g. lines drawn from the numbers to the appropriate regions of the diagram. Do not accept numbers written in more than one region.
5	(4S2) (5S1)	195,637 43,392	1 1	

6	(4G2c)	Diagram completed correctly as shown: Mirror line	1	Accept inaccurate drawing within a radius of 2 mm of the correct point, provided the intention is clear. Diagram need not be shaded. Diagram need not include edges drawn along the gridlines.
7	(6F2)	$\dfrac{4}{5} = \dfrac{8}{10} = \dfrac{16}{\mathbf{20}}$	1 1	
8	(5F10)	Numbers circled: 0.07　　　0.3	1	Accept alternative unambiguous indications, e.g. numbers ticked or underlined.
9	(5M9a)	Award **TWO** marks for answer of 37p Award **ONE** mark for evidence of an appropriate method, e.g. $190 \div 2 = 95$; $132 - 95$ OR $190 \div 10 = 19$; $5 \times 19 = 95$; $132 - 95$	Up to 2	Answer need not be obtained for the award of **ONE** mark. Award **ONE** mark for answer of 0.37p OR £37p OR £37 as evidence of an appropriate method.
10	(3F2)	Award **TWO** marks for all shading as shown: Any four squares of the oblong, any eight segments of the circle and any two sections of the parallelogram. Award **ONE** mark for two diagrams correct.	Up to 2	Accept alternative unambiguous indications of parts shaded.
11	(5M9c)	Award **TWO** marks for answer of 50 Award **ONE** mark for evidence of an appropriate method, e.g. • 1.5 kg = 1,500 g; 1,500 ÷ 30	Up to 2	Answer need not be obtained for **ONE** mark. Units must be converted correctly for **ONE** mark.
12	(6A2)	76 30	1 1	
13	(6R1)	Award **TWO** marks for answer of 156 Award **ONE** mark for evidence of an appropriate method, e.g. • 180 ÷ 30 = 6; 4 × 6 = 24; 180 − 24 OR • 180 ÷ 30 = 6; 30 − 6 = 24; 24 × 6	Up to 2	Answer need not be obtained for the award of **ONE** mark.
14	(6C5)	18 AND 36 only	1	Numbers may be given in either order.
15	(5M5)	Award **TWO** marks for answer of 95°F Award **ONE** mark for evidence of an appropriate method, e.g. • 104 − 86 = 18; 18 ÷ 2 = 9; 9 + 86 OR • 104 − 86 = 18; 18 ÷ 2 = 9; 104 − 9 OR • 104 + 86 = 190; 190 ÷ 2	Up to 2	Answer need not be obtained for the award of **ONE** mark.
16	(6G4b) (6G4a)	$a = 40$ $b = 70$	1 1	If the answers to a and b are incorrect, award **ONE** mark if $a + b = 110°$
17	(6N2)	8,999,990 4,800,000	1 1	
18	(6C8)	200	1	
19	(6C8)	Award **THREE** marks for answer of £127.20 Award **TWO** marks for: • sight of £102 AND £10.20 AND £15.00 as all multiplication steps completed correctly OR • sight of 10,200p AND 1,020p AND 1,500p as all multiplication steps completed correctly. • evidence of an appropriate complete method with no more than one arithmetic error. Award **ONE** mark for evidence of an appropriate complete method.	Up to 3	Answer need not be obtained for the award of **ONE** mark. No marks are awarded if there is more than one misread or if the maths is simplified. Award **TWO** marks if an appropriate complete method with the misread number is followed through correctly. Award **ONE** mark for all multiplication steps completed correctly with the misread number OR evidence of an appropriate complete method with the misread number followed through correctly with no more than one arithmetic error.
20	(6P2)	(−15, −20)	1	

Set B, Paper 1: arithmetic

Question (Content domain)		Requirement	Mark
1	(3N2b)	1,053	1
2	(3C2)	442	1
3	(4C6b)	407	1
4	(3C1)	568	1
5	(3C2)	1,219	1
6	(3C7)	14	1
7	(5C2)	84,816	1
8	(3C1)	609	1
9	(3C7)	31	1
10	(4C7)	2,268	1
11	(3C7)	558	1
12	(5C6a)	3,200	1
13	(5C6b)	62,700	1
14	(5F8)	8.349	1
15	(5C7b)	175	1
16	(5F8)	44.398	1
17	(5F8)	188.97	1
18	(5C2)	230,329	1
19	(6C9)	36	1
20	(6F9a)	0.03	1
21	(4F8)	3.75	1
22	(4C6b)	110	1
23	(5C7a)	Award **TWO** marks for answer of 2,997. Award **ONE** mark for a formal method of long multiplication with no more than **ONE** arithmetic error. Working must be carried through to a final answer for **ONE** mark. **Do not** award any marks if the error is in the place value, e.g. the omission of the zero when multiplying by tens.	Up to 2
24	(4F4)	$1\frac{2}{9}$ or $\frac{11}{9}$. Accept equivalent fractions or **exact** decimal equivalent, e.g. 1.222. Do not accept rounded or truncated decimals.	1
25	(6R2)	810. Do not accept 810%	1
26	(6F9b)	64.5	1
27	(5F4)	$\frac{1}{3}$. Accept equivalent fractions or **exact** decimal equivalent, e.g. 0.333'.	1
28	(6C7b)	Award **TWO** marks for answer of 26. Working must be carried through to a final answer for **ONE** mark. Award **ONE** mark for a formal method of division with no more than **ONE** arithmetic error.	Up to 2
29	(6R2)	93	1
30	(6C7a)	Award **TWO** marks for answer of 313,698. Working must be carried through to a final answer for **ONE** mark. Award **ONE** mark for a formal method of long multiplication with no more than **ONE** arithmetic error. **Do not** award any marks if the error is in the place value, e.g. the omission of the zero when multiplying by tens.	Up to 2
31	(6F4)	$1\frac{11}{20}$ or $\frac{31}{20}$. Accept equivalent mixed numbers, fractions or **exact** decimal equivalent, e.g. 1.55. Do not accept rounded or truncated decimals.	1
32	(6C7b)	Award **TWO** marks for answer of 51. Working must be carried through to a final answer for **ONE** mark. Award **ONE** mark for a formal method of division with no more than **ONE** arithmetic error.	Up to 2
33	(6F5b)	$\frac{1}{7}$. Accept equivalent fractions or **exact** decimal equivalent, e.g. 0.142	1
34	(5F5)	36	1
35	(6F4)	$\frac{8}{15}$. Accept equivalent fractions or **exact** decimal equivalent e.g. 0.533' (accept any unambiguous indication of the recurring digit). Do not accept rounded or truncated decimals.	1
36	(6C9)	88	1

Set B, Paper 2: reasoning

Question (Content domain)		Requirement	Mark	Additional guidance
1	(4M4b)	Both clocks ticked: 07:50 and 19:50	1	Accept alternative unambiguous indications, e.g. circled.
2	(6N5)	6	1	Do not accept –6 or 6–
		–8	1	Do not accept 8–
3	(3C1)	Award **TWO** marks for numbers in order as shown: **61** 77 93 **109** 125 141 **157**. Award **ONE** mark for two numbers correct.	Up to 2	
4	(6A2)	▲ = 31. ★ = 24	1 / 1	If the answers to ▲ and ★ are incorrect, award **ONE** mark if ▲ + ★ = 55
5	(5F8)	0.079 0.509 0.89 3.001 4.8	1	
6	(4F10b)	Award **TWO** marks for answer of 1.70. Award **ONE** mark for evidence of an appropriate method, e.g. • 2.45 + 1.85 = 4.30; 6 – 4.30 • 6 – 2.45 = 3.55; 3.55 – 1.85	Up to 2	Award **ONE** mark for answer of 170 metres as evidence of an appropriate method. Answer need not be obtained for the award of **ONE** mark.
7	(4G4)	A and F	1	Letters may be in either order.
		B, D and E	1	Letters may be in any order.

©HarperCollins*Publishers* 2018

8	(6C8)	Award **TWO** marks for answer of 42p OR £0.42
		Award **ONE** mark for evidence of an appropriate method, e.g.
		• 50p + 20p + 2p + 2p = 74p
		  £2.00 – 74p = £1.26
		  £1.26 ÷ 3
			Up to 2	Award **ONE** mark for answer of £42 OR £42p OR 0.42p as evidence of an appropriate method.
				Answer need not be obtained for the award of **ONE** mark.

9	(5S1)	47 10:45	1 1	The answer is a time interval. The answer is a specific time.
10	(6C7a)	Award **TWO** marks for answer of 4,800 Award **ONE** mark for evidence of an appropriate method with no more than one arithmetic error.	Up to 2	Misreads are not allowed.
11	(5M8)	B	1	Accept 40
12	(4P2)	The triangle has moved 6 squares to the right and 3 squares down.	1	
13	(4M4c)	480 7,200	1 1	
14	(3C4)	Award **TWO** marks for answer of 24 Award **ONE** mark for evidence of an appropriate method, e.g. • 3.5 × 4 = 14; 14 – 8 = 6; 6 × 4	Up to 2	Answer need not be obtained for the award of **ONE** mark.
15	(5N4)	Numbers circled: 400, 5,000	1	Accept alternative unambiguous indications, e.g. ticked.
16	(5N4)	Award **TWO** marks for: 40,900; 4,100; 400 Award **ONE** mark for two boxes correct.	Up to 2	
17	(5M9c)	Award **TWO** marks for answer of 1.8 Award **ONE** mark for evidence of an appropriate method, e.g. • 1.5 × 12 = 18; 18 ÷ 10	Up to 2	Answer need not be obtained for the award of **ONE** mark.
18	(4G2a)	Award **TWO** marks for rhombus AND square ticked. Award **ONE** mark for: • rhombus AND square and not more than one incorrect shape ticked OR • one correct shape only ticked.	Up to 2	
19	(6G2a)	E	1	
20	(5F10)	Award **TWO** marks for answer of £12.90 Award **ONE** mark for evidence of an appropriate method, e.g. • £1.75 + £2.55 = £4.30; £4.30 × 3	Up to 2	Award **ONE** mark for answer of £1,290 OR £1,290p OR £12.9 as evidence of an appropriate method. Answer need not be obtained for the award of **ONE** mark.
21	(6C8)	An explanation that shows that 6,832 can be made by adding 427 to 16 × 427, e.g. • '6,832 + 427 = 17 × 427' • '17 × 427 is 427 more than 6,832' • 'Because this is the same as 16 × 427 = 6,832 so add one more 326 to get the answer' • 'You add 427 to 6,832 and your answer will be correct' • 'Because you can add 427 to the answer of 16 × 427' • '6,832 + 427'.	1	Do not accept explanation that simply calculates 427 × 17 = 7,259 Do not accept vague or incorrect explanations, e.g. • 'You could add another 427' • 'The difference between 16 and 17 is 1 so you add 427 and that is one more'.

Set B, Paper 3: reasoning

Question (Content domain)		Requirement	Mark	Additional guidance
1	(5C6b)	100	1	
2	(4C8)	74 × 5	1	
3	(3C8)	6	1	
4	(6N5/ 6S1)	10 –4	1 1	Do not accept –10 or 10– Do not accept 4–
5	(5S1) (4M4b)	The correct time circled: 14:25	1	Accept alternative unambiguous indications, e.g. 14:25 ticked. Accept 15:47 circled as well as 14:25, provided no other is circled.
6	(4F10b)	221.19	1	
7	(6P2)	Triangle with vertices at (1,3) AND (4,3) AND (1,6) drawn as shown: 	1	Accept very slight inaccuracies in drawing.

8	(6C5)	Award **TWO** marks for any three of the following numbers written in any order: 5, 10, 20, 40 Award **ONE** mark for two numbers correct.	Up to 2	
9	(5S1) (4S2)	6	1	Do not accept 360 minutes.
10	(6G5)	16	1	
11	(4M5)	518(ml) or 0.518(l)	1	Do not accept incorrect units.
12	(6G2a) (4G4)	Circled correctly:	1	Accept alternative unambiguous indications, e.g. ticked.
13	(4C8)	An explanation that shows Clara has four times as many biscuits as Harry, e.g. • 32×4 is 4 times as many as 16×2 • 128 is four times 32 • $128 \div 4 = 32$ • $128 \div 32 = 4$ • $32 \times 4 = 128$ • Clara buys twice as many packets of twice as many biscuits, so it's doubled twice • 32 is double 16 and 4 is double 2, so it's doubled twice • Harry buys half the amount of biscuits and each packet has half the number of biscuits, so he has $\frac{1}{4}$ of the amount.	1	Do not accept vague or incomplete explanations, e.g. • Clara buys more packets and there are more biscuits in each packet • Clara buys twice as many packets of twice as many biscuits • 32 is double 16 and 4 is double 2.
14	(6R1) (5M9a)	Award **TWO** marks for answer of £1.32 Award **ONE** mark for evidence of an appropriate method, e.g. • $£1.65 \times 4 = £6.60$; $£6.60 \div 5$	Up to 2	Award **ONE** mark for answer of £132p OR 13.2p as evidence of an appropriate method. Answer need not be obtained for the award of **ONE** mark.
15	(5C8b)	Award **TWO** marks for answer of 1,250 Award **ONE** mark for evidence of an appropriate method, e.g. • $350 \times 2 = 700$; $3,200 - 700 = 2,500$; $2,500 \div 2$	Up to 2	Answer need not be obtained for the award of **ONE** mark.
16	(3G2)	**H** is circled.	1	Accept alternative unambiguous indications, e.g. letter ticked.
17	(6F11) (6F3)	Award **TWO** marks for all rows circled correctly as shown: $1\frac{1}{5}$; $1\frac{3}{4}$; 1.5; 1.6 Award **ONE** mark for two or three correct.	Up to 2	Accept alternative unambiguous indications of the correct numbers, e.g. numbers ticked.
18	(5C5c) (5C5d)	Both numbers correct as shown: 9 + 11 OR 1 + 19	1	Numbers must be in correct order. **Do not** accept: $3^3 + 11$ or $1^2 + 19$
19	(5M7b) (5C7a)	Award **THREE** marks for answer of 18 Award **TWO** marks for: • 918 as evidence of 34×27 completed correctly **OR** • evidence of an appropriate method with no more than one arithmetic error. Award **ONE** mark for evidence of an appropriate method.	Up to 3	Answer need not be obtained for the award of **ONE** mark. A misread of a number may affect the award of marks. Award **TWO** marks for an appropriate method using the misread number followed through correctly to a final answer.
20	(4N6)	Award **TWO** marks for 11 AND 12 Award **ONE** mark for: • only one correct number and no incorrect number **OR** • 11 AND 12 AND not more than one incorrect number.	Up to 2	Award **ONE** mark for answer of 66 AND 72 AND no more than one incorrect number.
21	(6F4/6A3)	$\frac{1}{4}$ written in the first box	1	Accept equivalent fractions or **exact** decimal equivalent, e.g. 0.25
		$3\frac{1}{4}$ or $\frac{13}{4}$ written in the last box	1	Accept equivalent fractions or **exact** decimal equivalent, e.g. 3.25
22	(6A1)	Award **TWO** marks for answer of 19 Award **ONE** mark for evidence of an appropriate method, e.g. • 15 + 4 + 2 widths = 38 + 1 width 19 + 2 widths = 38 + 1 width 19 + 1 width = 38; 38 − 19	Up to 2	Answer need not be obtained for the award of **ONE** mark. Award **ONE** mark for a method which uses algebraic representation correctly.
23	(6M8b) (6R1)	Award **TWO** marks for answer of 32 Award **ONE** mark for evidence of an appropriate method, e.g. • $8 \times 8 \times 8 = 512$; $512 \div 8 = 64$; $64 \div 2$	Up to 2	Answer need not be obtained for the award of **ONE** mark.
24	(6A4)	Both numbers correct as shown: $b = 4 \times a - 2$	1	

Key Stage 2

Mathematics Set A

Paper 1: arithmetic

Name						
School						
Date of Birth	Day		Month		Year	

Instructions

Do not use a calculator to answer the questions in this test.

Questions and answers

You have **30 minutes** to complete this test.

Write your answer in the box provided for each question.

You should give all answers as a single value.

For questions expressed as mixed numbers or common fractions, you should give your answers as mixed numbers or common fractions.

If you cannot do a question, move on to the next one, then go back to it at the end if you have time.

If you finish before the end of the test, go back and check your answers.

Marks

The numbers under the boxes at the side of the page tell you the number of marks for each question.

Answers are worth one or two marks.

Long division and long multiplication questions are worth **TWO marks each**. You will get TWO marks for a correct answer; you may get **ONE mark** for showing a correct method.

1 23 + 3000 =

2 657 + 2,546 =

3 $\dfrac{4}{8} + \dfrac{7}{8} =$

4   345 ÷ 1 =

1 mark

5   465 − 80 =

1 mark

6   4.5 + 5.023 =

1 mark

7 _____ = 3,700 + 800

1 mark

8 7 × 26 =

1 mark

9 84 ÷ 12 =

1 mark

10 $235 \times 3 =$

11 $3{,}813 - 647 =$

12 $\dfrac{54}{80} - \dfrac{26}{80} =$

13

_____ − 200 = 2,089

14

72 + (48 ÷ 8) =

15

$\frac{2}{7} \times \frac{3}{6} =$

16 $50 \times 60 =$

17 $472 \div 8 =$

18 $0.08 \div 10 =$

19

$3,465 \times 1,000 =$

1 mark

20

Show your method

$19\overline{)608}$

2 marks

21

$8 - 2.64 =$

1 mark

22

$$
\begin{array}{r}
5\ 3\ 7\ 5 \\
\times \quad \quad 3\ 6 \\
\hline
\end{array}
$$

Show your method

2 marks

23

$$\frac{2}{5} - \frac{3}{10} =$$

1 mark

24

$$
\begin{array}{r}
5\ 2\ 7 \\
\times \quad 6\ 2 \\
\hline
\end{array}
$$

Show your method

2 marks

25 45.9 − 23.739 =

26 $\dfrac{1}{3} + \dfrac{1}{6} + \dfrac{1}{10} =$

27 $\dfrac{3}{8} \div 3 =$

28 $\dfrac{5}{6} \div 2 =$

29 35% of 520 =

30 $3\dfrac{2}{5} + \dfrac{7}{10} =$

31 8% of 700 =

1 mark

32 $\dfrac{4}{5} - \dfrac{2}{8} =$

1 mark

33 0.6 × 300 =

1 mark

34

12% × 2000 =

35

$1\frac{1}{2} \times 63 =$

36

4 7 | 1 8 3 3

Show your method

This is a blank page

This is a blank page

Key Stage 2

Mathematics
Set A

Paper 2: reasoning

Name						
School						
Date of Birth	Day		Month		Year	

Instructions

Do not use a calculator to answer the questions in this test.

Questions and answers

You have **40 minutes** to complete this test.

Follow the instructions carefully for each question.

If you need to do working out, use the space around the question.

Some questions have a method box. For these questions, you may get one mark for showing the correct method.

Show your method

If you cannot do a question, move on to the next one, then go back to it at the end if you have time.

If you finish before the end of the test, go back and check your answers.

Marks

The numbers under the boxes at the side of the page tell you the number of marks for each question.

1 Sara asked the children in Year 3 and Year 5 if they had a pet.

This graph shows the results.

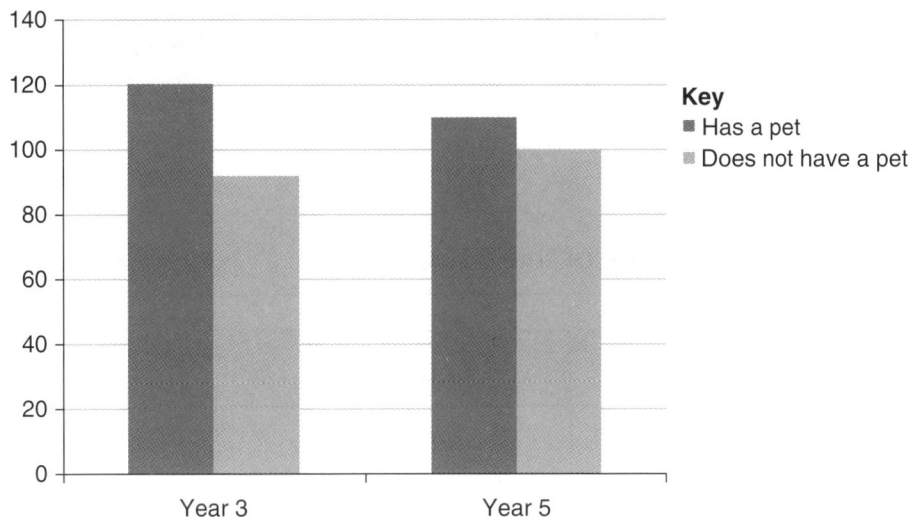

Key
- Has a pet
- Does not have a pet

Altogether, how many children do not have a pet?

1 mark

How many more Year 3 children than Year 5 children have a pet?

1 mark

2 Write the missing numbers to make this multiplication grid correct.

x		
8	72	40
	108	60

1 mark

3 Circle the number that is ten times greater than four hundred and three.

4,300   403   4,003   430   4,030

4 This table shows the lengths of three rivers.

River	Length in km
Nile	6,853
Zambezi	2,693
Loire	1,012

How much longer is the river Nile than the combined length of the other two rivers?

Show your method

km

5 Write the missing numbers.

72 months = ☐ years

96 hours   = ☐ days

77 days   = ☐ weeks

2 marks

6 Complete this table with the missing numbers.

One row has been done for you.

Number	1,000 more
2,800	3,800
26	
	8,054
	17,269

2 marks

7 At the start of May, there were 1,576 tennis balls in the sports shop.

During May,

- 9,854 more tennis balls were delivered

- 8,546 tennis balls were sold.

How many tennis balls were left in the shop at the end of May?

Show your method

2 marks

8 Tick two shapes that have $\frac{3}{4}$ shaded.

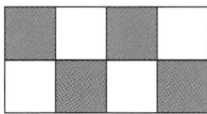

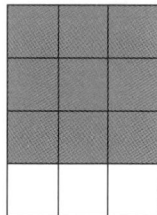

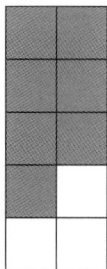

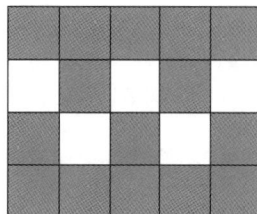

1 mark

9 Here is a rule for the time it takes to cook roast beef.

Cooking time = 30 minutes plus an extra 20 minutes for each kilogram

How many minutes will it take to cook a 4 kg joint of beef?

minutes

1 mark

What is the mass of a joint of beef that takes 90 minutes to cook?

kg

1 mark

10 Round 76,615

to the nearest 10:

to the nearest 100:

to the nearest 1,000:

2 marks

11 Here are diagrams of some 3-D shapes.

Tick each shape which has more vertices than faces.

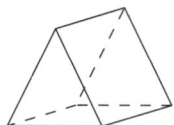

Triangular prism ☐

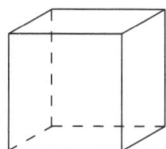

Cube ☐

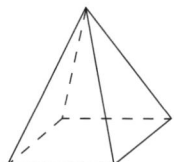

Square-based pyramid ☐

Triangular-based pyramid ☐

2 marks

12 Write these numbers in order of size, starting with the **smallest**.

2.5　　0.72　　2.134　　0.428

1 mark

13 Kate and Daniel buy some lollies.

Box of 12 lollies

£10.49

12 lollies

95p each

Kate buys a pack of 12 lollies for £10.49

Daniel buys 12 single lollies for 95p each.

How much more does Daniel pay than Kate?

Show your method

p

2 marks

14 At the end of a TV programme, the year it was made is given in Roman numerals.

MMXII

Write the year MMXII in figures.

1 mark

15 Amy poured some drinks at a party.

For every 4 drinks Amy poured, only 3 were drunk.

Altogether, 15 drinks were drunk.

How many drinks did Amy pour?

1 mark

16 Craig spins this spinner:

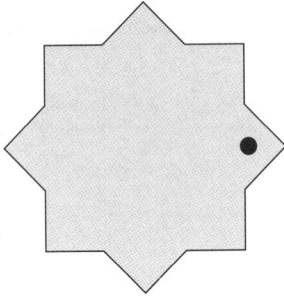

The spinner completes one-and-a-half turns before it stops.

Draw the new position of the dot on this spinner:

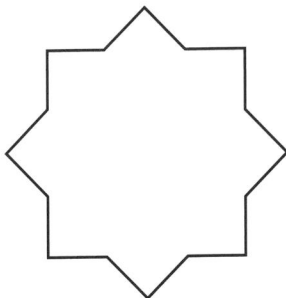

1 mark

17 Amina posts three parcels.

The postage costs the same for each parcel.

She pays with a £20 note.

Her change is £14.81

What is the cost of posting one parcel?

Show your method

£

2 marks

18 Ben has driven 20 kilometres. 10% of his journey has been completed.

Write the missing percentage.

Ben has driven 30 kilometres. [ ] % of his journey has been completed.

1 mark

19 The vertices of a quadrilateral have these coordinates.

(2, 5) (6, 2) (1, –7) (–6, 5)

One side of the quadrilateral has been drawn on the grid.

Complete the quadrilateral.

Use a ruler.

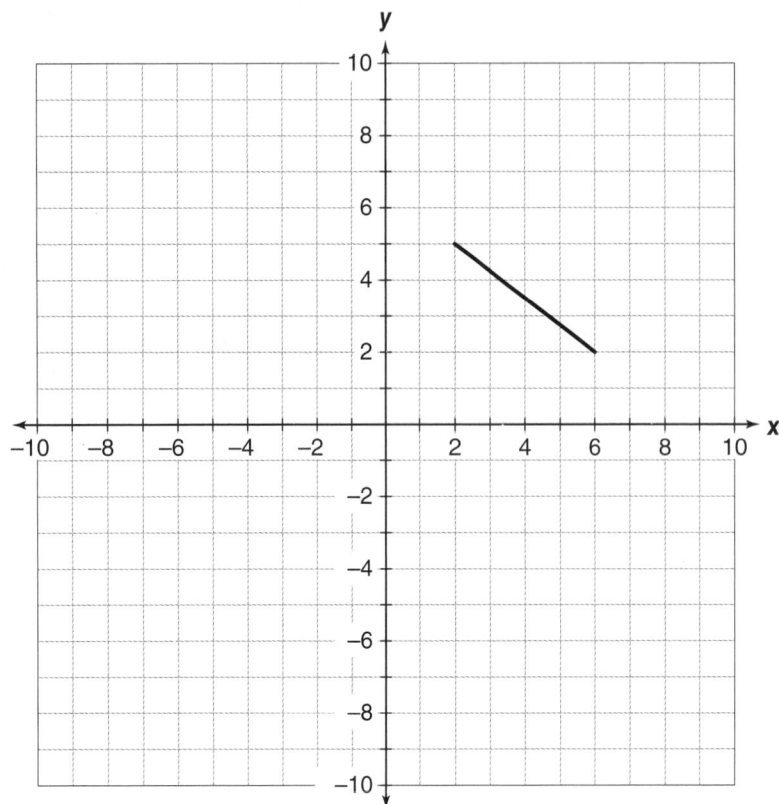

1 mark

20 Mohammed says

'0.4 is smaller than $\frac{4}{5}$,'

Explain why he is correct.

1 mark

21 On a map, 1 cm represents 50 km.

The distance between two towns is 775 km.

On the map, what is the measurement in cm between the two towns?

Show your method

cm

2 marks

22 In this circle, $\frac{1}{4}$ and $\frac{1}{8}$ are shaded.

What fraction of the whole circle is **not** shaded?

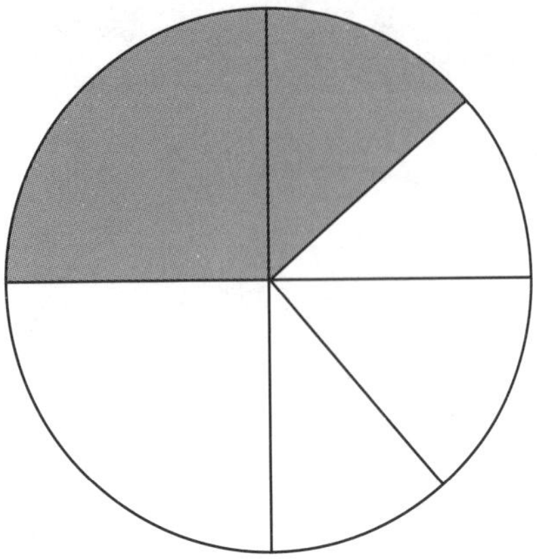

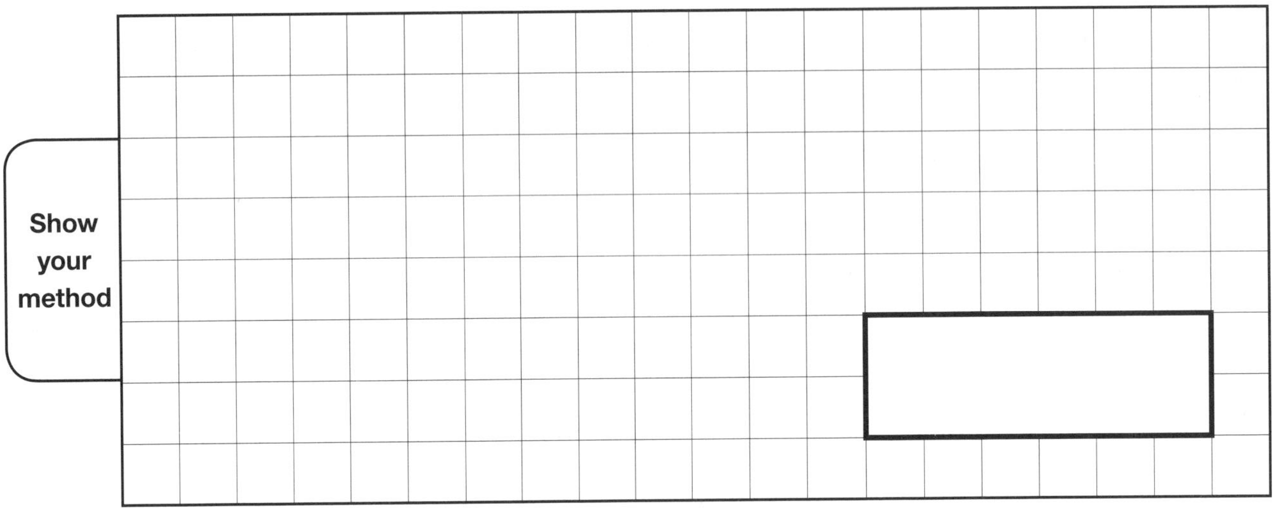

Show your method

2 marks

23 Here are two similar quadrilaterals.

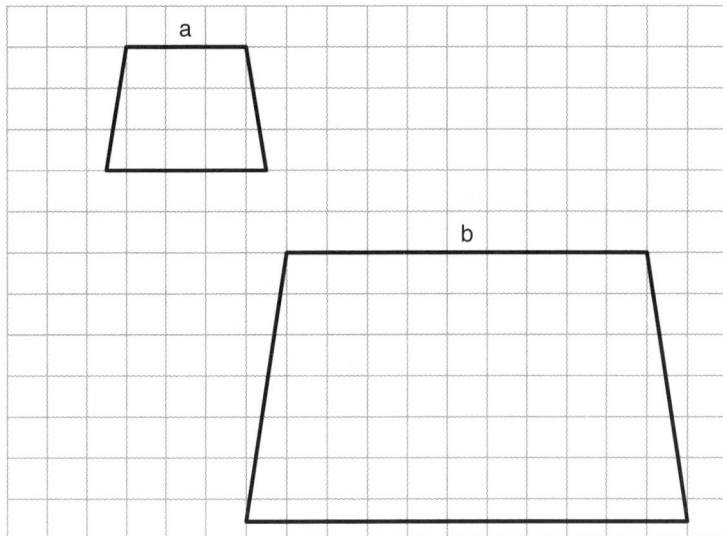

Write the ratio of side *a* to side *b*.

1 mark

This is a blank page

Mathematics
Set A

Paper 3: reasoning

Name						
School						
Date of Birth	Day		Month		Year	

Instructions

Do not use a calculator to answer the questions in this test.

Questions and answers

You have **40 minutes** to complete this test.

Follow the instructions carefully for each question.

If you need to do working out, use the space around the question.

Some questions have a method box. For these questions, you may get one mark for showing the correct method.

Show your method

If you cannot do a question, move on to the next one, then go back to it at the end if you have time.

If you finish before the end of the test, go back and check your answers.

Marks

The numbers under the boxes at the side of the page tell you the number of marks for each question.

1 Put these sports car prices in order, starting with the **lowest price**.

One has been done for you.

A — £147,250

B — £104,150

C — £140,500

D — £151,600

E — £87,500

	B			

1 mark

2 Ali puts these five numbers in their correct places on a number line.

709   698   703   771   638

Write the number closest to 700

1 mark

Write the number furthest from 700

1 mark

3 Write the three missing digits to make this addition correct.

$$
\begin{array}{r}
2 \quad 7 \quad \boxed{} \\
+\, 5 \quad \boxed{} \quad 2 \\
\hline
\boxed{} \quad 2 \quad 6
\end{array}
$$

2 marks

4 Write each number in its correct place on the diagram.

19   21   23   25

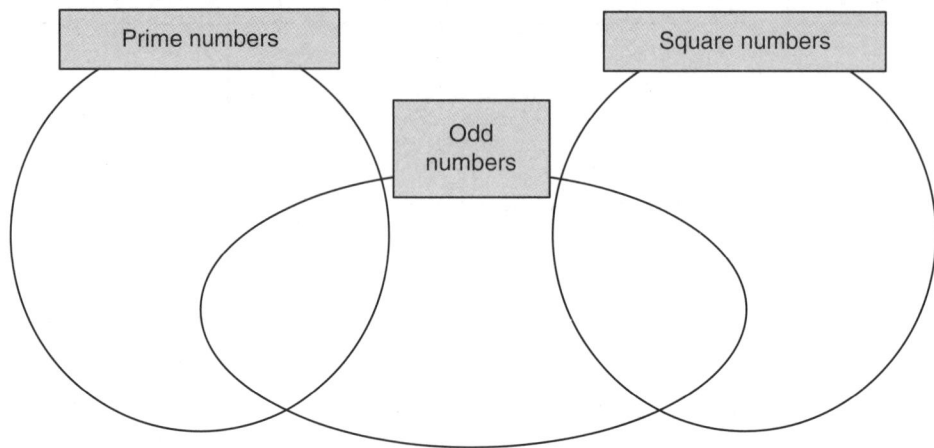

2 marks

5 This table shows the number of people living in various towns in England.

Town	Population
Harrogate	73,576
Tipton	42,407
Dunstable	30,184
Huntingdon	23,937
Middlesbrough	171,700

What is the **total** of the numbers of people living in Huntingdon and Middlesbrough?

Show your method

1 mark

What is the difference between the numbers of people living in Harrogate and in Dunstable?

Show your method

1 mark

6 This diagram shows a shaded shape inside a border of squares.

Draw the reflection of the shape in the mirror line.

Use a ruler.

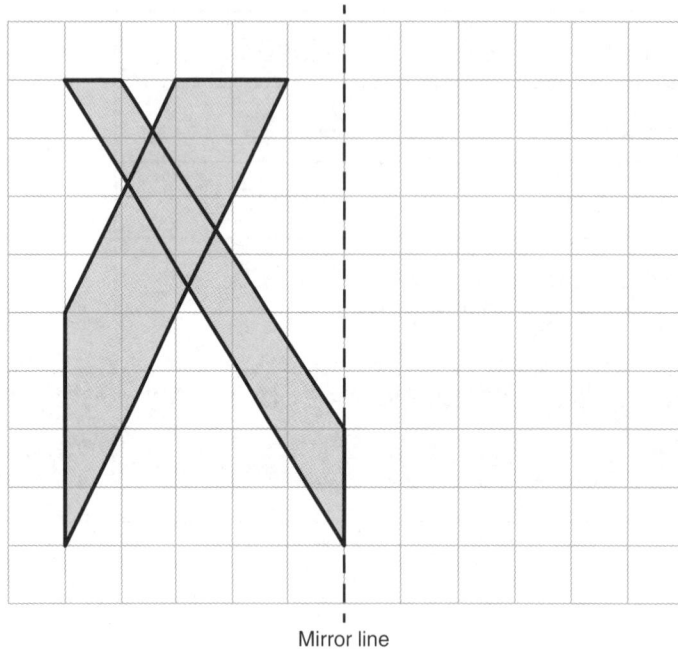

Mirror line

1 mark

7 Write the two missing values to make these equivalent fractions correct.

$$\frac{\boxed{}}{5} = \frac{8}{10} = \frac{16}{\boxed{}}$$

2 marks

8 Circle two numbers that add together to equal 0.37

0.07   0.33   0.3   0.7

1 mark

9 10 pens cost £1.90

5 pens and 1 ruler cost £1.32

What is the cost of 1 ruler?

Show your method

p

2 marks

10 Each diagram below is divided into equal sections.

Shade two-fifths of each diagram.

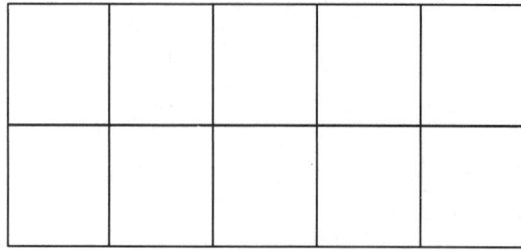

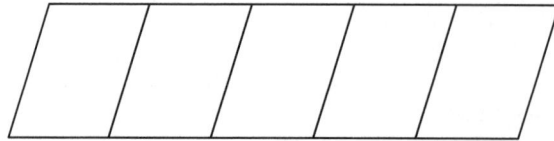

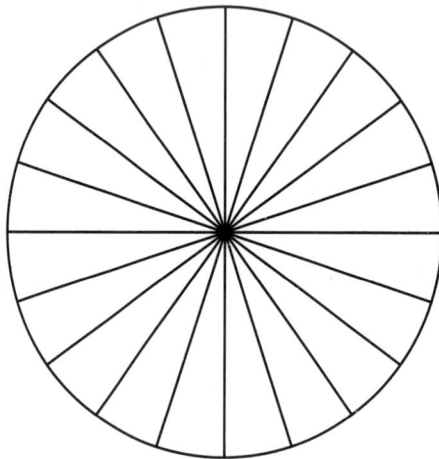

2 marks

11 A packet contains 1.5 kg of cereal.

Every day Maria eats 30 g of cereal.

How many days does the packet of cereal last?

Show your method

2 marks

12 $x = 34$

What is $2x + 8$?

1 mark

$3z + 20 = 110$

Work out the value of z.

1 mark

13 A stack of 30 identical books is 180 cm tall.

Ahmed takes four books off the top of the stack.

How tall is the stack now?

Show your method

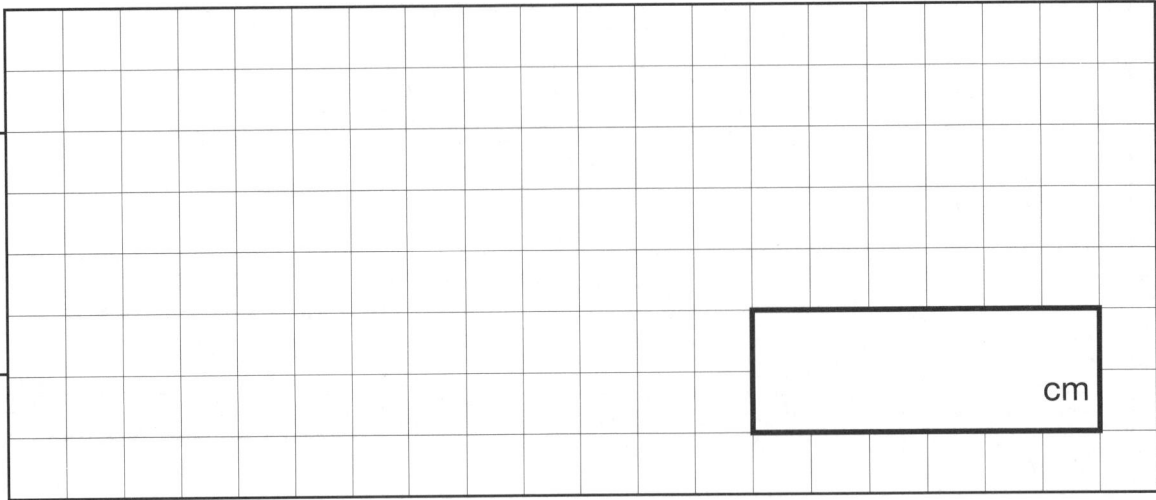

cm

2 marks

14 Write all the common multiples of 6 and 9 that are less than 50

1 mark

15 This thermometer shows temperatures in both °C and °F.

Work out what 35 °C is in °F.

Show your method

°F

2 marks

16 Calculate the size of angles *a* and *b* in this diagram.

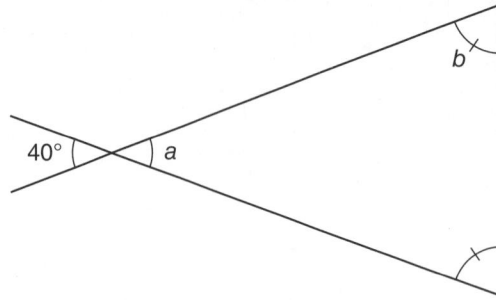

$a =$ [   ] $b =$ [   ]

17 Write the number that is ten less than nine million.

[   ]

Write the number that is two hundred thousand less than five million.

[   ]

18 Write the missing number.

$$500 \div \boxed{} = 2.5$$

1 mark

19 Miss Smith is making jam to sell at the school fair.

Raspberries cost £8.50 per kg.

Sugar costs 85p per kg.

10 glass jars cost £7.50

She uses 12 kg of raspberries and 12 kg of sugar to make 20 jars full of jam.

Calculate the total cost to make 20 jars full of jam.

Show your method

£

3 marks

20 Here are two triangles drawn on coordinate axes.

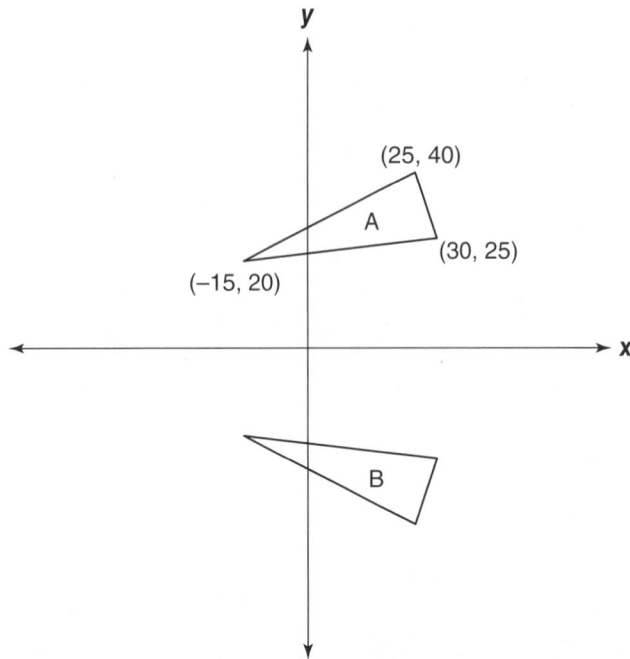

Triangle B is a reflection of triangle A in the x-axis.

Two of the new vertices of triangle B are (30, −25) and (25, −40).

What are the coordinates of the third vertex of triangle B?

1 mark

This is a blank page

This is a blank page

Mathematics
Set B

Paper 1: arithmetic

Name	
School	

Date of Birth	Day		Month		Year	

Instructions

Do not use a calculator to answer the questions in this test.

Questions and answers

You have **30 minutes** to complete this test.

Write your answer in the box provided for each question.

You should give all answers as a single value.

For questions expressed as mixed numbers or common fractions, you should give your answers as mixed numbers or common fractions.

If you cannot do a question, move on to the next one, then go back to it at the end if you have time.

If you finish before the end of the test, go back and check your answers.

Marks

The numbers under the boxes at the side of the page tell you the number of marks for each question.

Answers are worth one or two marks.

Long division and long multiplication questions are worth **TWO marks each**. You will get TWO marks for a correct answer; you may get **ONE mark** for showing a correct method.

1 953 + 100 =

2 35 + 407 =

3 407 ÷ 1 =

4 577 − 9 =

1 mark

5 _____ = 864 + 355

1 mark

6 98 ÷ 7 =

1 mark

7   78,987 + 5,829 =

8   _____ = 659 − 50

9   124 ÷ 4 =

10 $756 \times 3 =$

11 $62 \times 9 =$

12 $40 \times 80 =$

13 $100 \times 627 =$

1 mark

14 $6.009 + 2.34 =$

1 mark

15 $525 \div 3 =$

1 mark

16 12.97 + 31.428 =

1 mark

17 258.37 − 69.4 =

1 mark

18 243,327 − 12,998 =

1 mark

19 $4^2 + 20 =$

20 $0.3 \div 10 =$

21 $6 - 2.25 =$

22 $1{,}210 \div 11 =$

23

Show your method

$$\begin{array}{r} 8\ 1 \\ \times\ 3\ 7 \\ \hline \end{array}$$

2 marks

24 $\dfrac{4}{9} + \dfrac{7}{9} =$

1 mark

25 30% of 2,700 =

26 15 × 4.3 =

27 $\frac{2}{5} - \frac{1}{15} =$

28

$$3\ 4\ \lfloor 8\ 8\ 4$$

Show your method

2 marks

29

15% of 620 =

1 mark

30

$$
\begin{array}{r}
7\ 4\ 6\ 9 \\
\times\quad\ \ 4\ 2 \\
\hline
\end{array}
$$

Show your method

2 marks

31 $1\frac{7}{10} - \frac{3}{20} =$

1 mark

32

$2\ 3\ |\ 1\ 1\ 7\ 3$

Show your method

2 marks

33 $\frac{4}{7} \div 4 =$

1 mark

34

$$\frac{3}{10} \times 120 =$$

1 mark

35

$$1\frac{1}{5} - \frac{2}{3} =$$

1 mark

36

$$80 + 64 \div 8 =$$

1 mark

This is a blank page

This is a blank page

Mathematics
Set B

Paper 2: reasoning

Name	
School	
Date of Birth	Day Month Year

Instructions

Do not use a calculator to answer the questions in this test.

Questions and answers

You have **40 minutes** to complete this test.

Follow the instructions carefully for each question.

If you need to do working out, use the space around the question.

Some questions have a method box. For these questions, you may get one mark for showing the correct method.

Show your method

If you cannot do a question, move on to the next one, then go back to it at the end if you have time.

If you finish before the end of the test, go back and check your answers.

Marks

The numbers under the boxes at the side of the page tell you the number of marks for each question.

1 A clock shows this time twice a day.

Tick the two digital clocks that show this time.

07:50   08:50   19:50   10:50   20:50

1 mark

2 This table shows the temperature at 9 am on three days in December.

2nd December	10th December	20th December
+4 °C	–2 °C	+1 °C

What is the difference between the temperature on 2nd December and the temperature on 10th December?

°C

1 mark

On 30th December the temperature was 9 degrees lower than on 20th December.

What was the temperature on 30th December?

°C

1 mark

3 The numbers in this sequence increase by 16 each time.

Write the missing numbers.

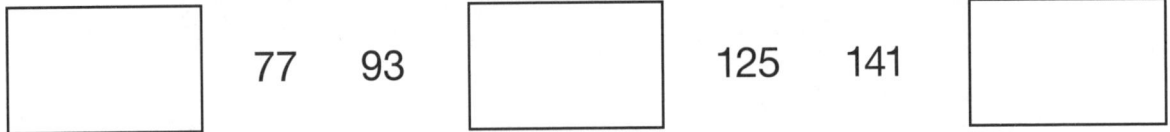

[ ]   77   93   [ ]   125   141   [ ]

2 marks

4 Each shape stands for a number.

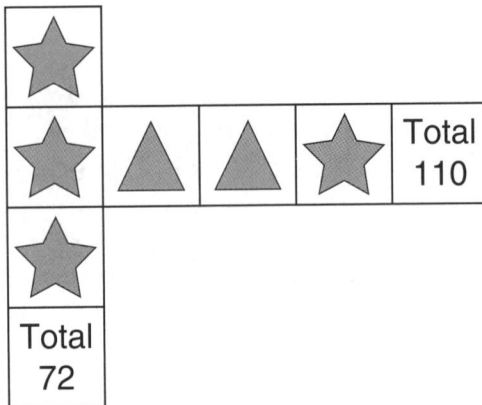

★
★ ▲ ▲ ★ | Total 110
★
Total 72

Work out the value of each shape.

▲ = [ ]

★ = [ ]

2 marks

5 Write these numbers in order, starting with the **smallest**.

0.89   0.509   4.8   0.079   3.001

1 mark

6 Jackie cuts 6 metres of rope into three pieces.

The length of the first piece is 2.45 metres.

The length of the second piece is 1.85 metres.

Work out the length of the third piece.

Show your method

m

2 marks

7 Here are five angles.

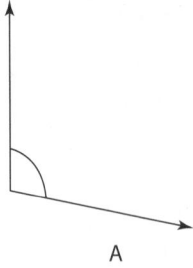

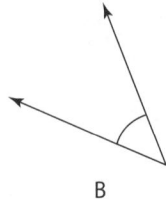

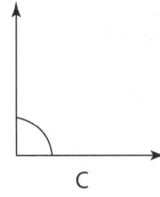

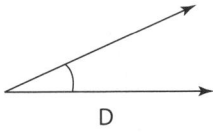

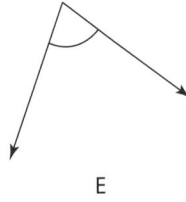

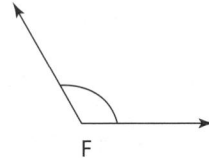

Write the letters of the angles that are obtuse.

Write the letters of the angles that are acute.

8 Asha buys three packets of crisps.

She pays with a £2 coin.

This is her change.

What is the cost of one packet of crisps?

Show your method										

p

2 marks

9 Here is part of the bus timetable from Greenvale to Riverford.

Greenvale	09:51	10:02	10:17	10:34
Winchley	09:59	10:10	10:24	10:41
New Harper	10:19	10:30	10:45	11:02
Mountsford	10:24	10:35	10:52	11:09
Riverford	10:34	10:45	11:04	11:21

How many minutes does it take the 10:17 bus from Greenvale to reach Riverford?

minutes

1 mark

Mr Jones is at New Harper at 10:21

What is the earliest time he can reach Riverford on the bus?

1 mark

10 A sweet shop orders 12 boxes of chocolates.

Each box contains 8 bags of chocolates.

Each bag contains 50 chocolates.

How many chocolates does the shop order in total?

Show your method

2 marks

11 Which of the following cuboids has the greatest volume?

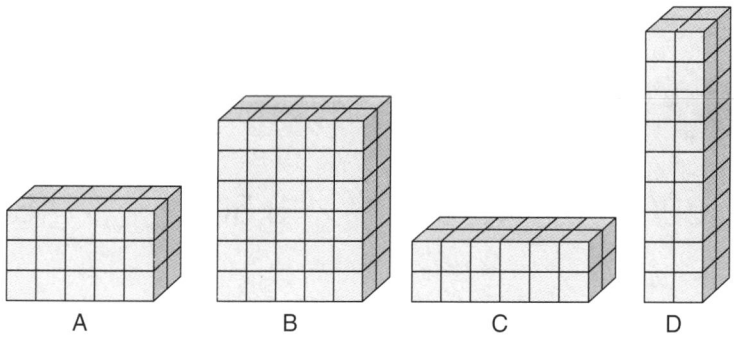

A   B   C   D

1 mark

12 A triangle is translated from position A to position B.

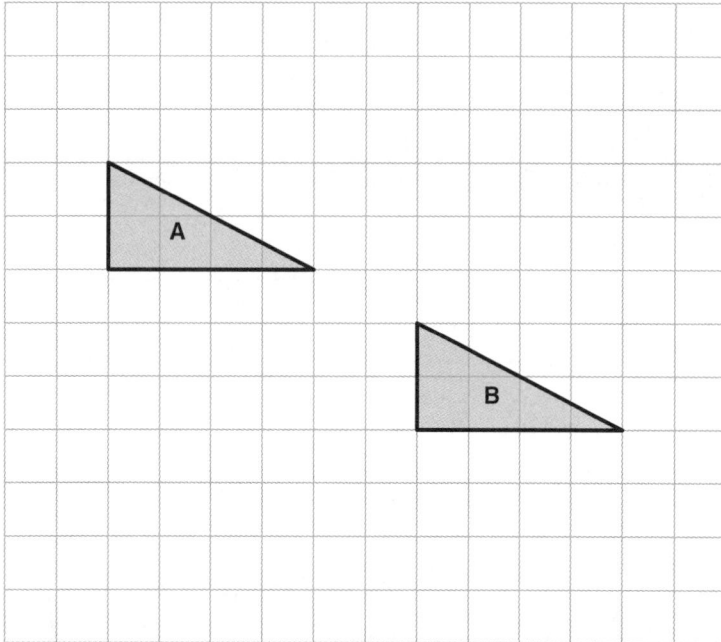

Complete the sentence.

The triangle has moved [   ] squares to the right and [   ]

squares down.

1 mark

13 Complete each sentence using a number from the list below.

| 1,200 | 2,400 | 600 | 7,200 | 3,600 | 480 |

There are [   ] minutes in eight hours.

1 mark

There are [   ] seconds in two hours.

1 mark

14 Lara chooses a number less than 30

She divides it by 4 and then adds 8

She then divides this result by 4

Her answer is 3.5

What was the number she started with?

Show your method

15 Circle two numbers that multiply together to equal 2 million.

400   4,000   5,000   50,000

16 Complete this table by rounding the numbers to the nearest hundred.

	Rounded to the nearest hundred
40,901	
4,080.5	
403.08	

2 marks

17 12 small bricks have the same mass as 10 large bricks.

The mass of one small brick is 1.5 kg.

What is the mass of one large brick?

Show your method

kg

2 marks

18 The diagonals of this quadrilateral cross at right angles.

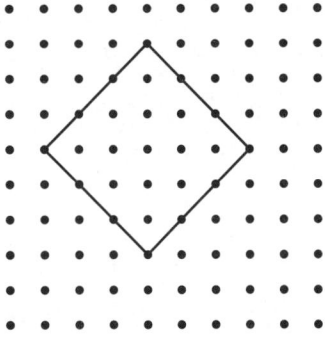

Tick all the quadrilaterals that have diagonals which cross at right angles.

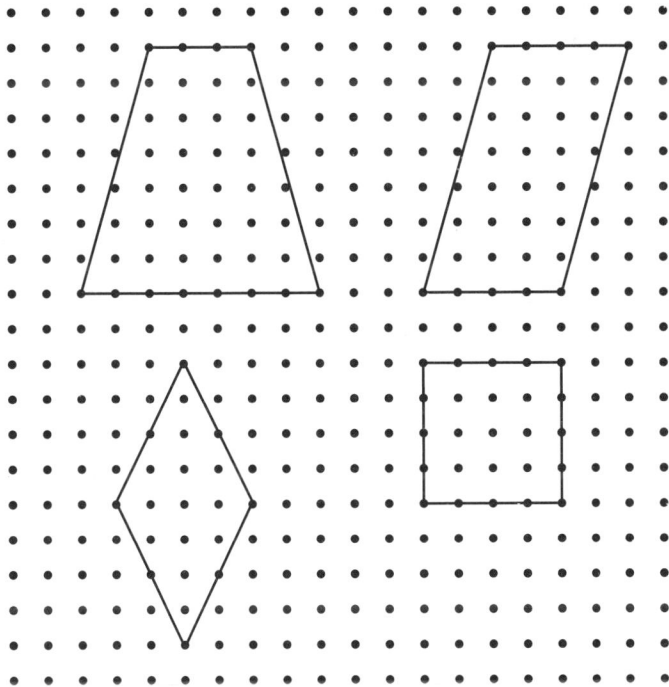

2 marks

19 Here are five quadrilaterals on a square grid.

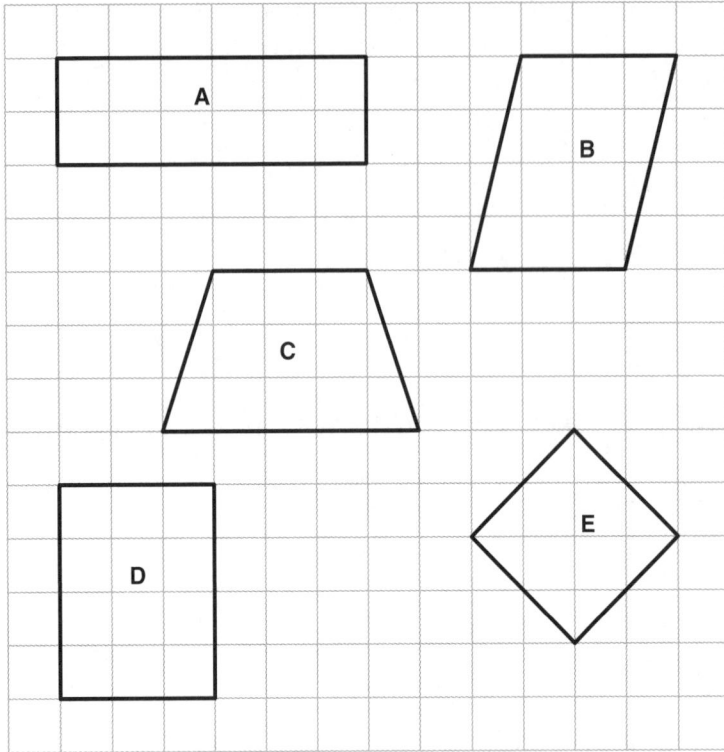

Four of the quadrilaterals have the same area.

Which quadrilateral has a different area?

1 mark

20 Lara had some money.

She spent £1.75 on a bus ticket.

She spent £2.55 on her lunch.

She has two-thirds of her money left.

How much money did Lara have to start with?

Show your method

£

2 marks

21 $6,832 \div 16 = 427$

Explain how you can use this fact to find the answer to 17×427

1 mark

Key Stage 2

Mathematics
Set B

Paper 3: reasoning

Name						
School						
Date of Birth	Day		Month		Year	

Instructions

Do not use a calculator to answer the questions in this test.

Questions and answers

You have **40 minutes** to complete this test.

Follow the instructions carefully for each question.

If you need to do working out, use the space around the question.

Some questions have a method box. For these questions, you may get one mark for showing the correct method.

Show your method

If you cannot do a question, move on to the next one, then go back to it at the end if you have time.

If you finish before the end of the test, go back and check your answers.

Marks

The numbers under the boxes at the side of the page tell you the number of marks for each question.

1 Write the missing number to make this division correct.

$15 \div \boxed{} = 0.15$

1 mark

2 Cody uses these digit cards.

5 4 7

He makes a 2-digit number and a 1-digit number.

He multiplies them together.

His answer is a multiple of 10

What could Cody's multiplication be?

$\boxed{}$ $\boxed{}$ × $\boxed{}$

1 mark

3 A group of friends earns £114 by doing odd jobs.

They share the money equally.

Each friend gets £19

How many friends are in the group?

$\boxed{}$

1 mark

4 This graph shows the temperature in °C from 2 am to 3 pm on a cold day.

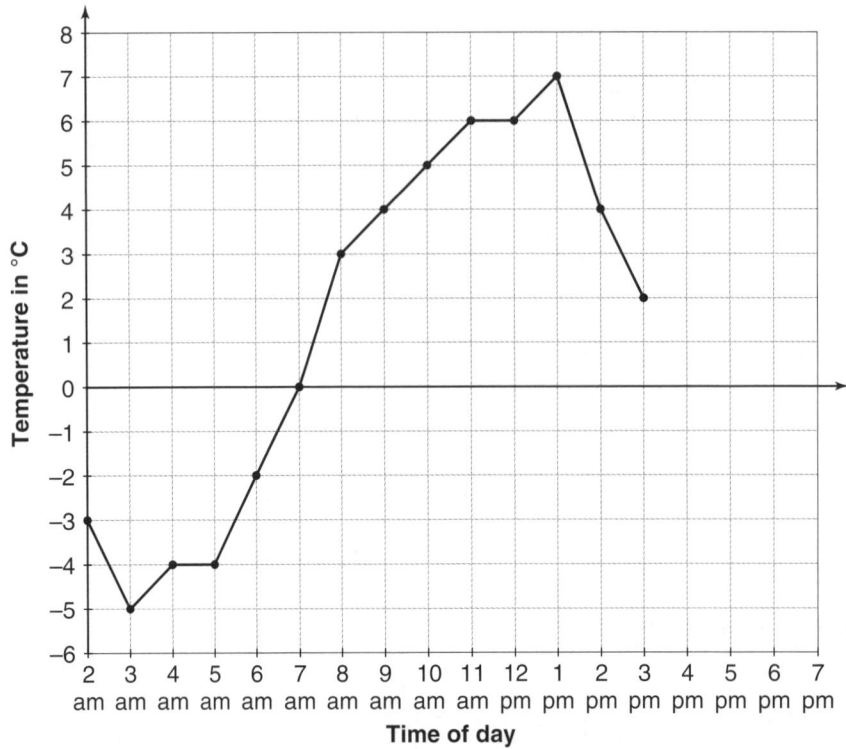

Time of day

How many degrees warmer was it at 12 pm than at 4 am?

°C

1 mark

At 7 pm the temperature was 8 degrees lower than at 2 pm.

What was the temperature at 7 pm?

°C

1 mark

5 Sue wants to travel to London by plane.

She needs to arrive in London by 4:30 pm.

Circle the latest flight that Sue can take.

Flight from Edinburgh	Arrives in London
12:15	13:30
13:15	14:30
14:25	15:47
15:25	16:40
17:15	18:25
18:40	19:55
20:05	21:40

1 mark

6 Charlie is saving up for a new phone.

He needs to save £250

So far he has saved £28.81

How much more does he need to save?

1 mark

7 Here is a triangle drawn on a coordinate grid.

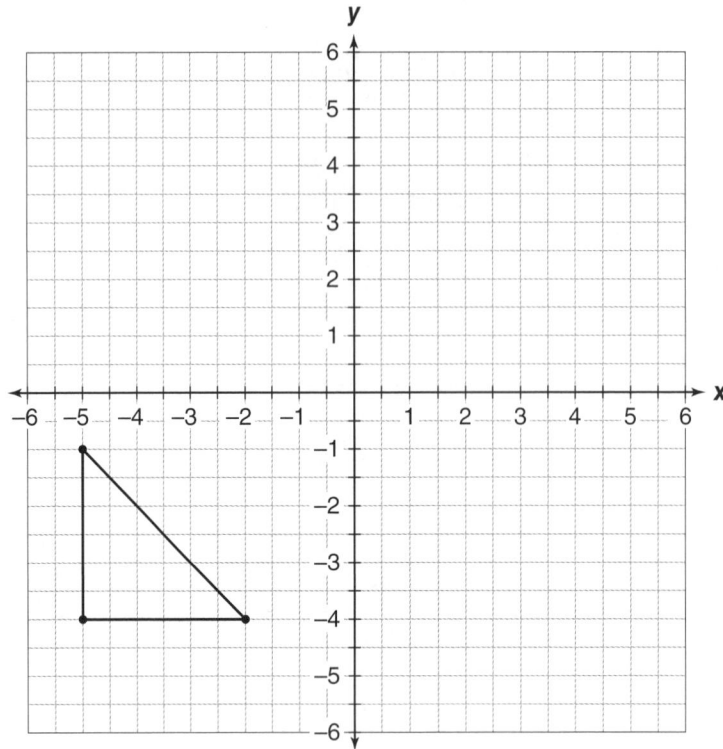

The triangle is translated 6 right and 7 up.

Draw the triangle in its new position.

1 mark

8 Write three factors of 40 that are not factors of 16

2 marks

9 Here is the timetable for fitness sessions at the leisure centre.

	Monday	Tuesday	Wednesday	Thursday	Friday
9:00 – 10:30	Aquacise	Zumba	Spinning	Aquacise	Spinning
10:30 – 11:30	Spinning	Aquacise	Zumba	Zumba	Aquacise
11:30 – 12:00	Zumba	Aquacise	Spinning	Aquacise	Zumba

What is the **total** number of hours for **Aquacise** classes on this timetable?

1 mark

10 A car steering wheel has a diameter of 32 cm.

What is the radius of the steering wheel?

cm

11 A bottle contains 768 millilitres of juice.

Jane pours out a quarter of a litre.

How much juice is left?

ml

12 Circle the hexagon which has exactly two acute angles.

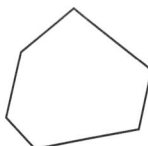

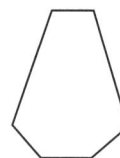

---

13 Clara buys 4 packs of chocolate biscuits.

Harry buys 2 packs of vanilla biscuits.

Packet of 32 chocolate biscuits   Packet of 16 vanilla biscuits

Clara says,

'I have four times as many biscuits as Harry.'

Explain why Clara is correct.

1 mark

14 5 pineapples cost the same as 4 coconuts.

One coconut costs £1.65

How much does one pineapple cost?

Show your method

£

2 marks

15 There are 3,200 marbles in a box.

Ivan and Sam take 350 marbles each.

Clara and Harry share the rest of the marbles equally.

How many marbles does Clara get?

Show your method

cm

2 marks

16 Look at the letters below.

Circle the letter below that has both parallel and perpendicular lines.

H   J   T   X   C

1 mark

17 In each box, circle the number that is greater.

$1\frac{1}{5}$	1.1
$1\frac{3}{4}$	1.3
$1\frac{1}{20}$	1.5
$1\frac{1}{2}$	1.6

2 marks

18 A square number and a prime number have a total of 20

What are the two numbers?

$$\boxed{} + \boxed{} = 20$$

square number   prime number

1 mark

19 A square of paper measures 30 cm by 30 cm.

A rectangular piece of paper is 4 cm **longer** and 3 cm **narrower** than the square piece.

What is the **difference in area** between the two pieces?

Show your method

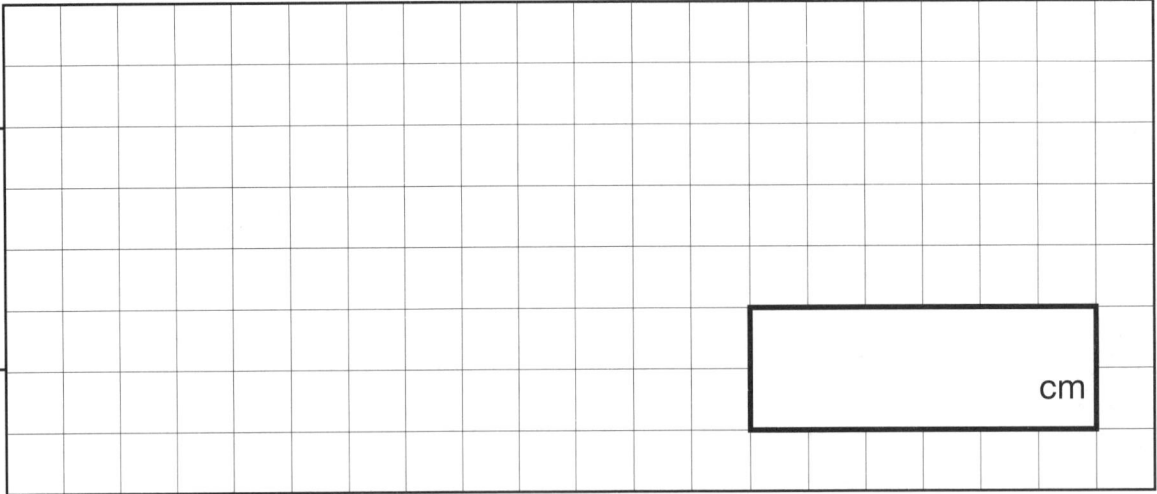

cm

3 marks

20 Harry thinks of a whole number.

He multiplies it by 6

He rounds his answer to the nearest 10

The result is 70

Write all the possible numbers that Harry could have started with.

2 marks

21 The numbers in this sequence increase by the same amount each time.

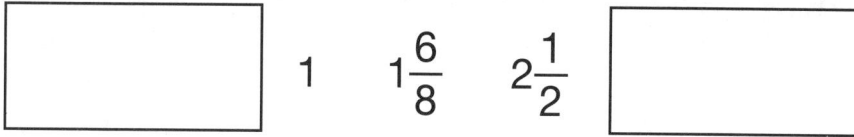

$$\boxed{} \quad 1 \quad 1\frac{6}{8} \quad 2\frac{1}{2} \quad \boxed{}$$

22 In this diagram, the shaded rectangles are all of equal width (w).

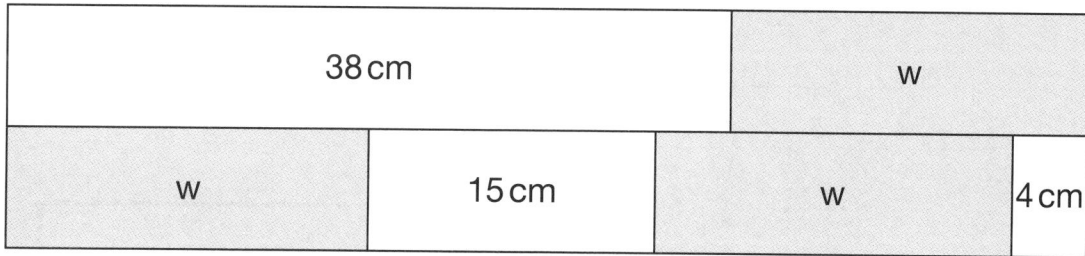

38 cm		w
w	15 cm	w	4 cm

Show your method

$$\boxed{} \text{ cm}$$

23 Cube A and cuboid B have the same volume.

Calculate the missing length on cuboid B.

Show your method

cm

2 marks

24 Here is a pattern of number pairs.

a	b
2	6
3	10
4	14
5	18

Complete the rule for the number pattern.

b = ☐ × a − ☐

1 mark

This is a blank page

This is a blank page